AF869074

Dark Matter is Nonexistent

Eugene J. Laviolette

Copyright © 2019 Eugene J. Laviolette

All rights reserved. No part of this book may be reproduced or transmitted in any form or by any means, electronic or mechanical, including photocopying, recording or by any information storage and retrieval system, without permission in writing from the publisher.

First Edition Print, 2019

Dark Matter Revealed Press - Bullhead City, AZ
ISBN: 978-0-578-54079-5
Dark Matter is Nonexistent | Eugene J. Laviolette
Available Formats: eBook | Paperback distribution

DARK MATTER IS NONEXISTENT
REVEALED BY MATHEMATICAL ERROR

LAVIOLETTEIAN RADIUS
PROJECT SOLEDAD

UNDERSTANDING LAVIOLETTEIAN SPHERICAL GEOMETRY

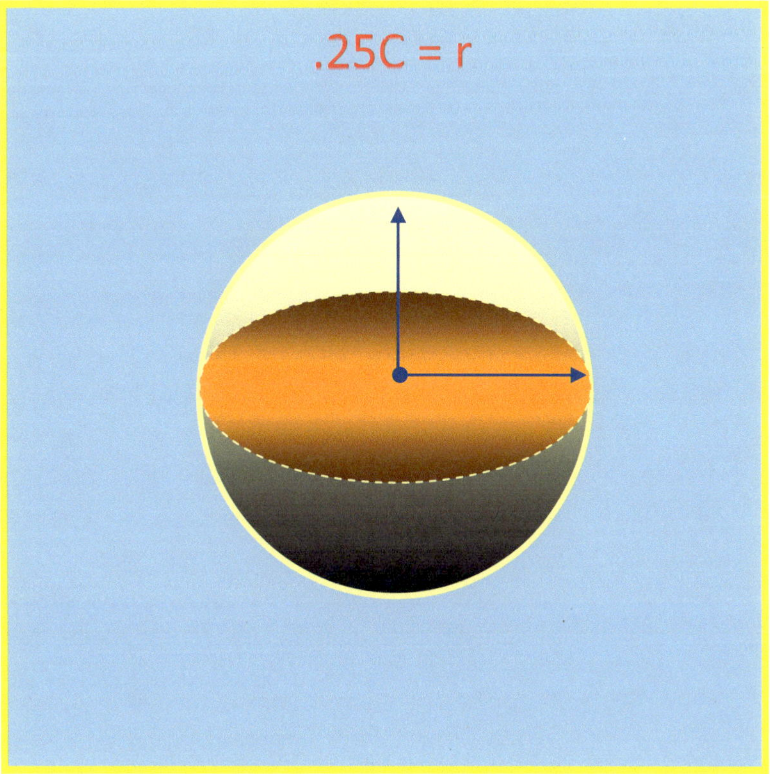

SPHERICAL SURFACE AREA
FROM THEORY TO THEOREM

By Eugene J. Laviolette

UNDERSTANDING LAVIOLETTEIAN SPHERICAL GEOMETRY

Photos from NASA Goddard Flight Center.

Pictures above reveal a planet created from a nebula disk via the accretion process.

PURPOSE: To prove Procedural conversion of Contemporary Second-Dimensional calculations to Third-Dimensional expression relevant to the sphere reflects a loss of **23.36985 %** of Surface Area **on the Second-Dimensional plane** because total Surface Area of **four Great Circles** converts circles to squared units inaccurately, relevant to expressing surface area of a Third-Dimensional Sphere, as well as, an accurate percentage amount of mass. **Third-Dimensional surface area and matter distribution is affected.**

ACCRETION PROCESS: It is not known with certainty how planets are formed. The prevailing theory is that they are formed during the collapse of a nebula into a thin disk of gas and dust. A protostar forms at the core, surrounded by a rotating protoplanetary disk. Through accretion (a process of sticky collision) dust particles in the disk steadily accumulate mass to form ever-larger bodies. Local concentrations of mass known as planetesimals form, and these accelerate the accretion process by drawing in additional material by their gravitational attraction. These concentrations become ever denser until they collapse inward under gravity to form protoplanets.

A planet's defining physical characteristic is that it is massive enough for the force of its own gravity to dominate over the electromagnetic forces thus binding its physical structure leading to a state of equilibrium but is not massive enough to cause thermonuclear fusion.

Orbital clearing: The defining dynamic characteristic of a planet is that it has ***cleared its neighborhood***. A planet that has cleared its neighborhood has accumulated enough mass to gather up or sweep away all the planetesimals in its orbit.

All celestial **stars** and **planetary-mass objects** massive enough for the force of its own gravity to dominate **become rounded under its own gravity** binding its physical structure, leading to a state of equilibrium. This effectively means that they are **spherical or spheroidal**. Up to a certain mass, an object can be irregular in shape, but beyond that point, which varies depending on the chemical makeup of the object, **gravity begins to pull an object towards its own center of mass until the object collapses into a sphere.**

The contemporary Second Dimensional Plane when utilized to reveal Surface Area of a **Third-Dimensional** Sphere via a **formula** must provide SUFFICIENT **Second-Dimensional** SURFACE COVERAGE to account for natures **Accretion Factor** if it were activated to reveal settled matter of a **Third-Dimensional** Sphere from a hypothetical Second-Dimensional Plane.

CLARIFICATION STATEMENT:

Any reference of **Third-Dimensional from Second-Dimensional** stated herein means: String material was placed on a flat surface having **Length and Width** then reshaped to view a **Third-Dimensional figure or shape** without adding or removing any material covering the **Second-Dimensional Plane Superior Circular figure.**

Third-Dimensional figures can be viewed by arcing and or assembling Second-Dimensional figures without loss or gain of surface area expression. Six squares create a cube. Two circles close the ends of the arced rectangle to create a cylinder. There are many combinations that can be created from the Second-Dimensional plane to be seen as Third-Dimensional figures. **However, viewing and calculating Third-Dimensional Spherical Surface area requires a different procedure.**

LAVIOLETTEIAN Geometry agrees with contemporary surface area for polygons and circles, but does not agree with surface area for the contemporary sphere expressed on **the Euclidean Second-Dimensional plane.**

The **Third-Dimensional** sphere must also be assembled from the Second-Dimensional plane. Two hemispheres **ARCED** from the Second-Dimensional plane **CREATES** the **Third-Dimensional Sphere**. Lavioletteian Superior Circles contain the proportionate amount of material needed to create hemispheres yielding the **desired diameter of a sphere**. **Therefore it is imperative that the diameter of a circular construction on the Second-Dimensional plane be equal to the 180 degree arc of the hemisphere.**

After the Second-Dimensional figure has been covered with cords of string matter material it can be arced into hemispheres returning to its original Third-Dimensional Spherical Great Circle Diameter. Two hemispheres equal a complete Sphere.

One cannot calculate surface area of a **Third-Dimensional** sphere until the proper circle has been drafted on the Second-Dimensional plane. Once the proper circle is constructed you can calculate surface area of a **Third-Dimensional** sphere on **the Second-Dimensional plane** utilizing Lavioletteian formula as follows: 2 pi r^2 x .900317. **r = .25% of any Spherical Great Circle.**

D = 6.28318" **divided by (½ pi) 1.570795 equals** **D = 4"**

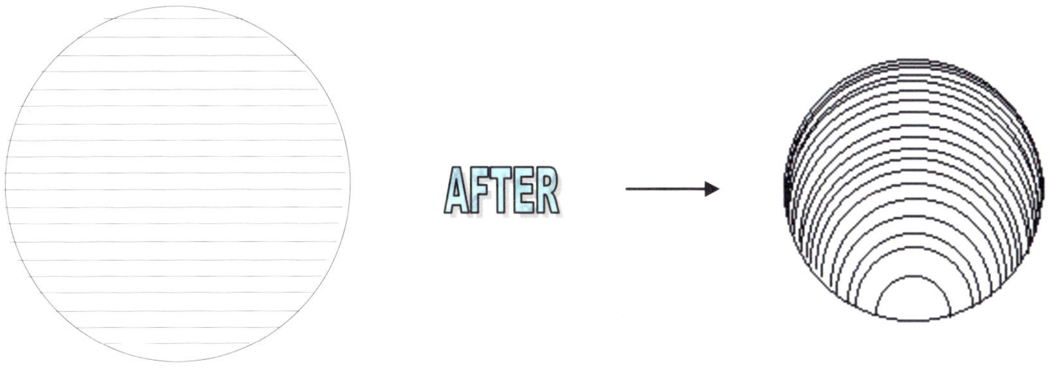

SUPERIOR CIRCLE **HEMISPHERE**

CONVERTED BY ARCING CORDS

Why is it incorrect to calculate spherical surface area from the following?

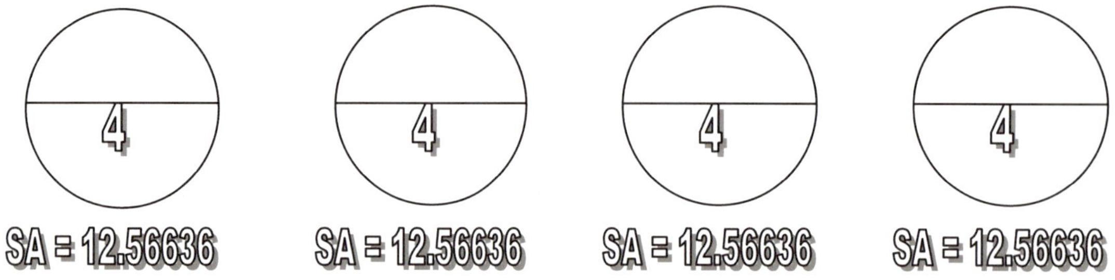

It is incorrect because it will understate true surface area of the Third-Dimensional sphere and **does not account for all the matter needed to be transferred from the Second-Dimensional plane** evident by the additional Lavioletteian 23.36985 % of missing matter now being utilized.

Laviolette's Superior Circles imply there is more matter and more spherical surface area. **WHY?**

NEW CONCEPT:

The fact is a sphere cannot be assembled or arced from the Second Dimensional plane and maintain its proper Great Circle Diameter without creating depth due to the fact Second-Dimensional circumference must be decreased as the arced hemisphere forms a Third-Dimensional figure. The following explanation will reveal more matter to properly express surface area of a sphere and explain why **it appears** we have 23 % of missing matter.

Let us first begin by understanding the concept of **squaring the circle**. The areas of this square and this circle are equal. In 1882, it was proven that this figure cannot be constructed in a finite number of steps with an idealized compass and straightedge.

<u>The circle in squaring the circle is equal to one Great Circle of any Contemporary Sphere.</u>

$\sqrt{3.14159} = 1.7724531$

1.7724531
x 1.7724531
3.1415899 = SA

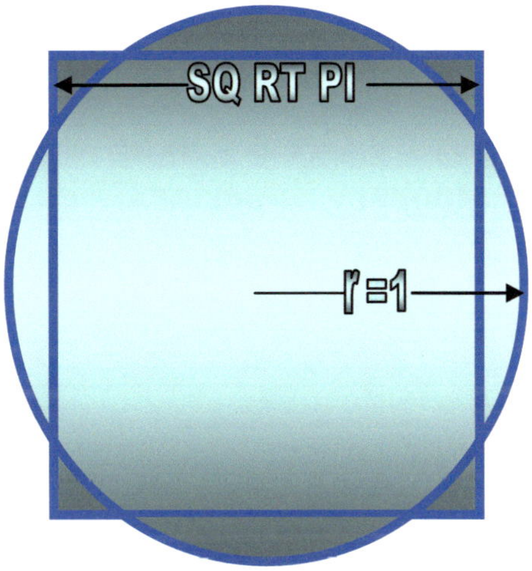

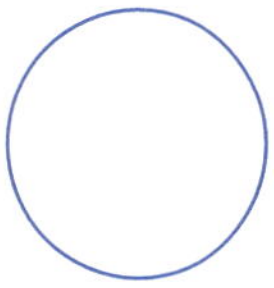

D = 2

πr^2
1 x 1 = 1
x 3.14159
3.14159 = SA

Surface Area of the circle is also equal to its D^2 x .7853975
2 x 2 = 4 x .7853975 = 3.14159

What is the linear measurement (lm) of string inside the square if string diameter is .075?
1.7724531 / .075 = 23.632708 x 1.7724531 = 41.887866 inches.

If you coil the 41.887866 inches of string on a Second-Dimensional plane the diameter of the circle will be 2 inches and the entire circle will be covered with string matter. NOTICE THE STRING MATTER WHEN PLACED ON THE SECOND-DIMENSIONAL PLANE REVEALS A DEPTH FACTOR. We will refer to this later in the report.

Perimeter divided by circumference yields the factor to calculate equal surface area for any square or circle if one or the other is known. 7.0898124 / 6.28318 = 1.1283796.

If the perimeter is known: 7.0898124 / 1.1283796 = 6.2831802 = 6.28318 = circumference.
If the circumference is known: 6.28318 x 1.1283796 = 7.0898121 = perimeter.
The answer is $\pm$ .0000003.

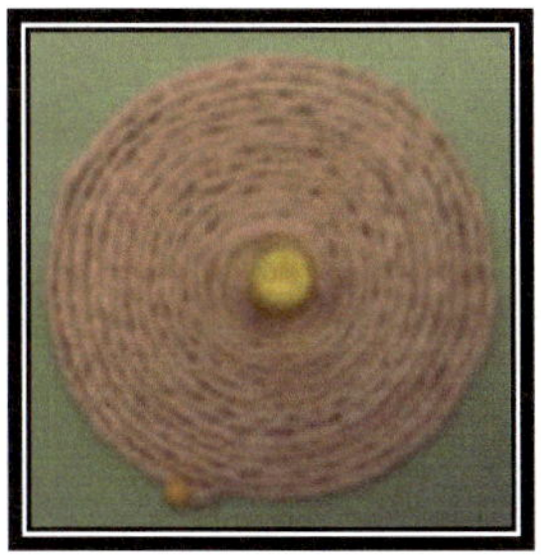

COILED STRING MATTER

CORD STRING MATTER

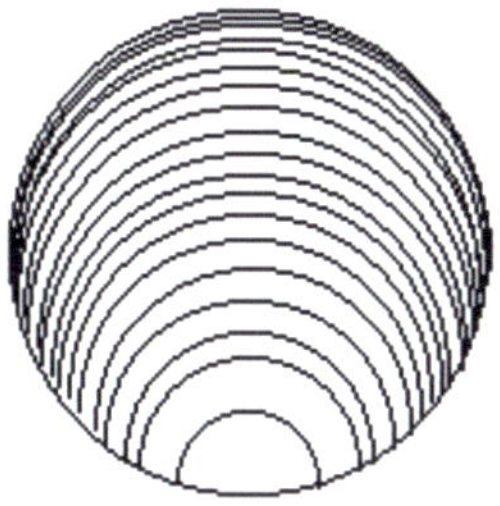

HEMISPHERE

Before we venture into Laviolettеian calculations let us first understand that a Laviolettеian Superior Circle represents the **CORD** material or matter needed to arc a Third-Dimensional Hemisphere from a Second-Dimensional plane.

As you know material from **two Contemporary Great Circles** would be needed to arc a Contemporary Hemisphere from a **Contemporary Superior Circle**. Let us demonstrate by using a Settled Spherical Great Circle of 2 inches.

1 Great Circle + 1 Great Circle

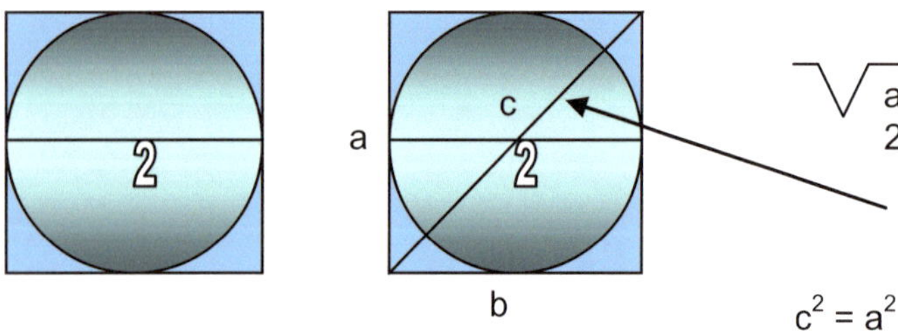

$\sqrt{a^2 + b^2} = \mathbf{c}$

2 x 2 + 2 x 2

4 + 4 **c**

8 = 2.8284271 = c

The root of 8 = c

$c^2 = a^2 + b^2$ Pythagorean Theorem

41.887866" lm + 41.887866" lm = 83.775732" lm. The circle from squaring the circle is equal to onc contemporary Great Circle of any contemporary sphere. The hypotenuse of the Great Circle **square** is equal to the diameter of a **Contemporary Superior Circle** which contains the string material or matter of two contemporary Great Circles. **.25 %** of the Spherical Great Circle = RADIUS for the Laviolettеian Superior Circle.

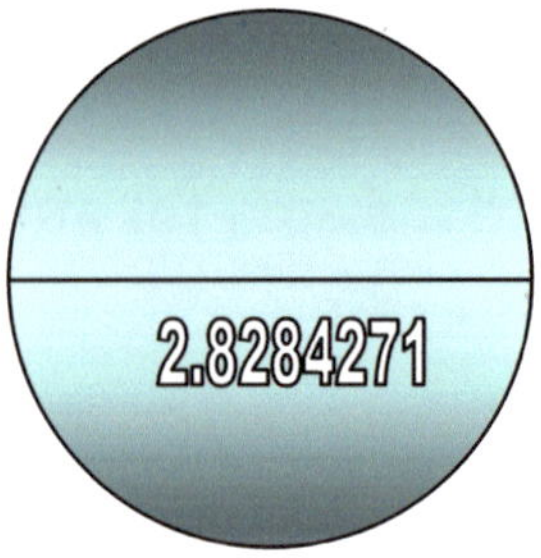

3.14159 / 2.8284271 = 1.1107198

2.8284271 / 3.14159 = .900317

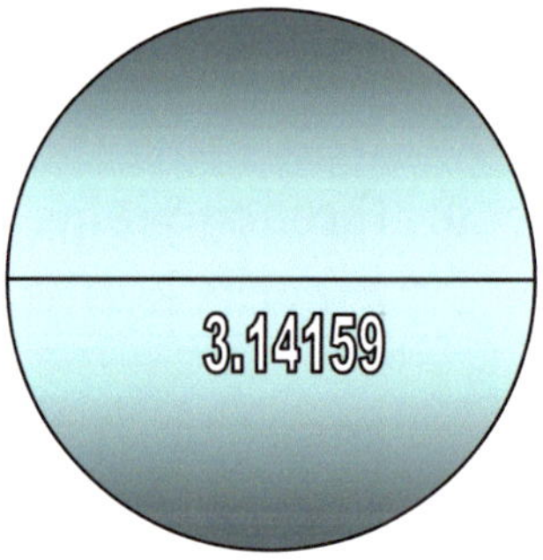

Contemporary Superior Circle

Laviolettеian Superior Circle

The diameter of the Laviolettеian Superior Circle is created from ½ of a Spherical Great Circle's circumference.

SA		SA
1.4142135		**1.5707950**
x 1.4142135		**x 1.5707950**
1.9999998		**2.4673969**
x 3.1415900		**x 3.1415900**
6.2831793		**7.7515494**
x 2	**πr^2 x 2 Superior Circles**	**x 2**
12.566358		**15.503098**

15.503098 / 12.566358 = 1.2336985

The Lavioletteian Surface Area yields 23.36985 % more surface area on the Second-Dimensional Plane.

The Contemporary *Superior Circle's* answer is the same as 4 Great Circles utilizing contemporary $4\pi r^2$.

As you can see below the **Contemporary Superior Circle**, consisting of 2 Contemporary Great Circles, is only **90.0317 %** of a Lavioletteian Superior Circle and its **CORDS** can only arc back to a Great Circle diameter of **1.8006341 inch instead of the required 2 inch Spherical Diameter.**

Contemporary SUPERIOR CIRCLE diameter 2.8284271 inch divided by (½ pi) 1.570795 equals a 1.8006341 inch HEMISPHERE Great Circle Diameter.

Laviolettian SUPERIOR CIRCLE diameter 3.14159 inch divided by (½ pi) 1.570795 equals a 2 inch HEMISPHERE Great Circle Diameter as required.

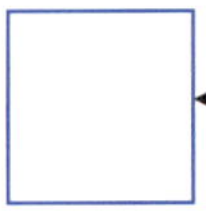

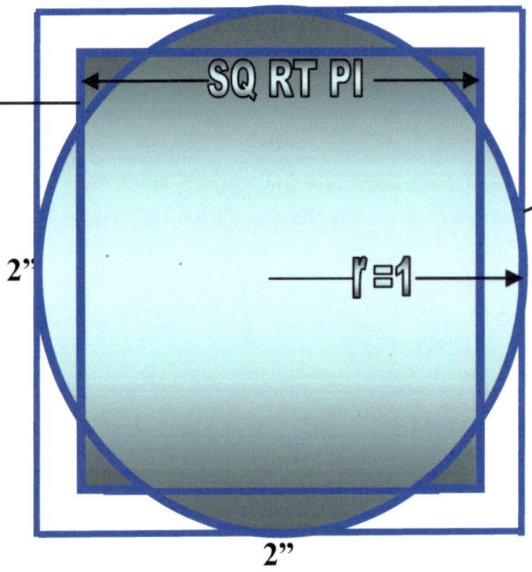

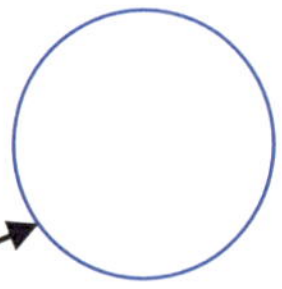

$\sqrt{3.14159}$ = 1.7724531

1.7724531
x 1.7724531
3.1415899 = SA

D = 2

πr^2
1 x 1 = 1
x 3.14159
3.14159 = SA

Lavioletteian Superior Circle arcs back to a Great Circle of 2 inches as required.

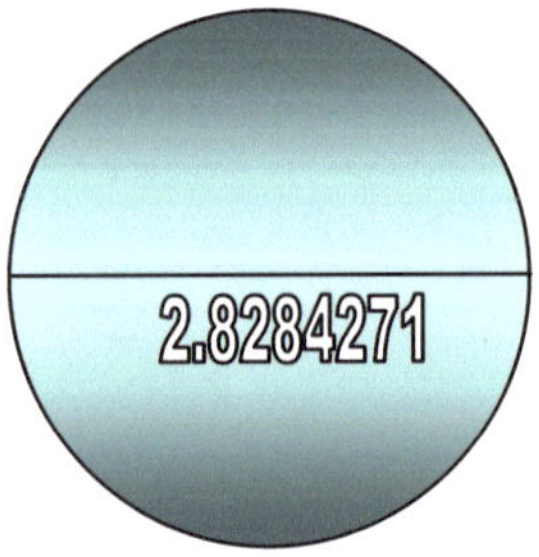

Contemporary Superior Circle

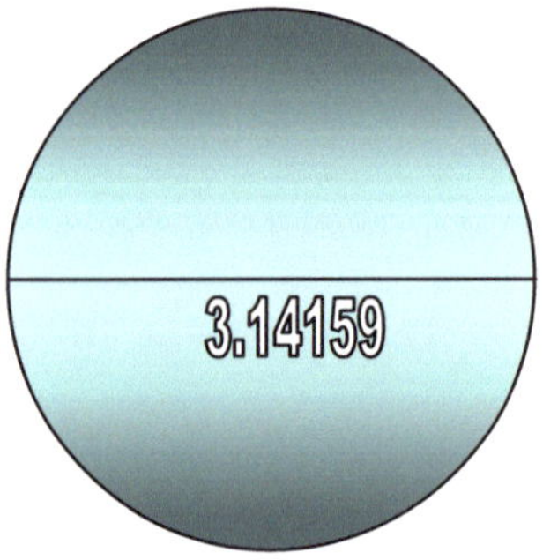

Lavioletteian Superior Circle

Contemporary Superior Circle		Lavioletteian Superior Circle
SA		**SA**
1.4142135		**1.5707950**
x 1.4142135		**x 1.5707950**
1.9999998		**2.4673969**
x 3.1415900		**x 3.1415900**
6.2831793		**7.7515494**
x 2	**πr^2 x 2 Superior Circles**	**x 2**
12.566358	**15.503098 / 12.566358 = 1.2336985**	**15.503098**
D = 3.14159	divided by (½ pi) 1.570795 equals	D = 2

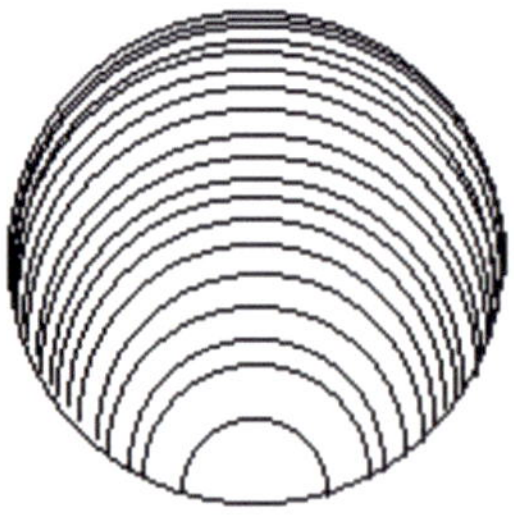

SUPERIOR CIRCLE

HEMISPHERE

Now that we have some understanding of the Laviolettеian concept we will run through the procedure from start to finish and prove **more** spherical surface area**, more** matter, **more** mass, **more** weight, **more** source strength and **more** force as we reveal the Dark Matter Mystery. We will calculate from a **Spherical Great Circle** diameter of 2.528 inch utilizing Laviolette's Second-Dimensional Superior Circles concept and **formula of $2\pi r^{2}$.**

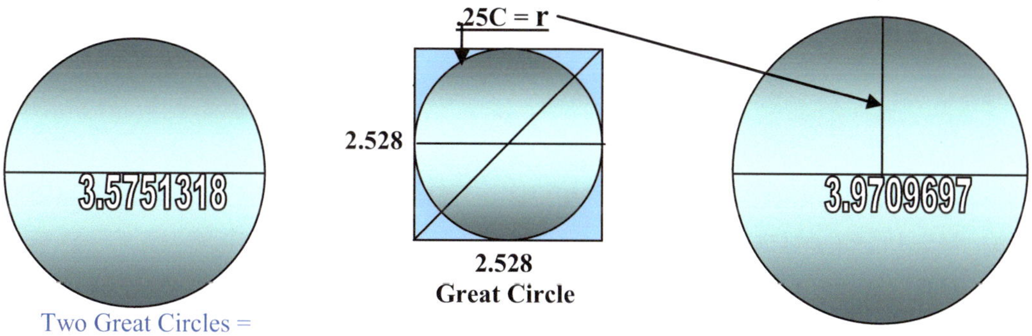

Two Great Circles =
Contemporary Superior Circle

Laviolettеian Superior Circle

The diameter of the Laviolettеian Superior Circle is created from ½ of a Spherical Great Circle's circumference.

SA		SA
1.7875659		1.9854848
x 1.7875659		x 1.9854848
3.1953918	**Reference page 16**	3.9421498
x 3.1415900		x 3.1415900
10.03861		12.384618
x 2	πr^2 **x 2 Superior Circles**	x 2
20.07722		24.769236

24.769236 / 20.07722 = 1.2336984

As you can see the Laviolettеian Surface Area yields 23.36985 % $\pm$ .0000001% more surface area.

Calculate **linear measurement** of string for both circles if string diameter is **#18** Twisted Mason Line.

Contemporary Superior Circle **Laviolettеian Superior Circle**

170 x 2 = 340 inches ⟶ x 1.2336985 = ⟶ 419.45749 inches

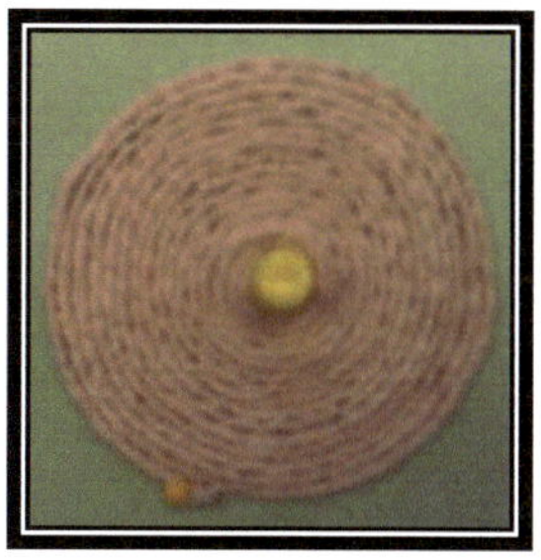

What is the length of linear measurement for the additional 23.36985 % of linear measurement?

The length is **419.45749 - 340** = 79.45749 inches.

Coil the 419.45749 inches of string or linear measurement matter on the surface of a **Third-Dimensional** sphere having a diameter of 2.528 **inches**. Note your observations. The sphere can be a tennis ball.

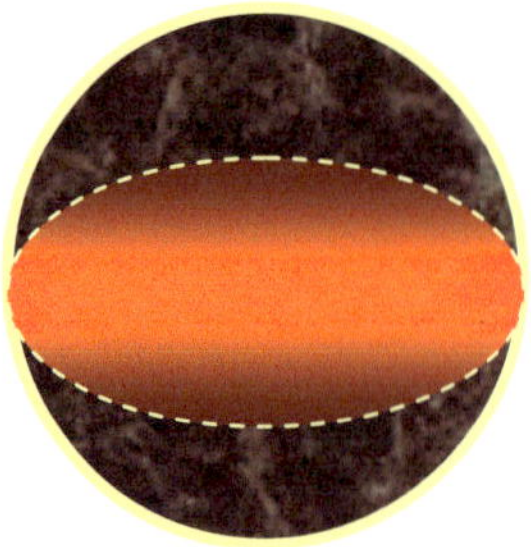

After the sphere was completely covered with the coiled linear measurement there was a piece leftover that measured 41.81276 ins. 419.45749 - 41.81276 = 377.64473 inches.

I concluded that the contemporary measurement of 340 inches is an insufficient amount to cover the entire sphere.

What was the total amount of linear measurement needed to cover the surface area of the sphere?

79.45749 for the additional 23.36985 % measure.
- 41.81276 piece leftover that measured 41.81276 inches.
37.64473

37.64473
+ **340.00000**
377.64473 inches was needed to cover the sphere.

What was the additional percent of linear measurement added to the contemporary measurement?
377.64473 / 340 = 1.1107197. The additional amount was 11.07197 %. ± .0000001
What was the percent of the additional linear measurement allocated to covering surface area of the sphere?
37.64473 / 79.45749 = .4737719 = 47.37719%

What was the percent of additional linear measurement leftover after covering surface area of the sphere?
1. - .4737719 = .5262281 = 52.62281%

What is the matter percentage of contemporary linear measurement compared to the actual linear measurement covering the sphere?
340. / 377.64473 = .900317 = 90.0317%

What is the matter percentage of actual linear measurement, compared to the contemporary linear measurement, covering the sphere?
377.64473 / 340 = 1.1107197 = 111.07197%

What happened to the 41.81276 inches of linear measurement or matter that was leftover?
It was absorbed internally as accumulated thickness on the Third-Dimensional plane when applied from the Second-Dimensional plane which altered diameter as it expressed Third-Dimensional form in the same way altering perimeter affects surface area when a square is altered to resemble a rectangle without removing any matter material on the Second-Dimensional plane and /or if material was utilized and equal perimeter was a maintained factor.

What is the matter percentage and amount of total Lavioletteian linear measurement allocated to the ball of the sphere?

1. - .900317 = .099683%.

419.45749 inches x .099683% = 41.81278 inches.

Surfaces Of Linear Emphatically Demand Accurate Dimensions.

Henceforth: Project Soledad.

Purpose: Proving contemporary Spherical Surface Area calculations understate Spherical Mass by 23.36985 % due to the fact the calculations are derived from the Perimeter linear measurement of a Squared geometrical shape rather than the Circumference of a Circular geometric shape. Thus, the **formulas are different**, as well as, the **total results and size of the shapes** on the Second Dimensional plane.

Laviolетteian Geometry reveals truth of the aforementioned statement. However, how do we prove the fact and reveal the missing mass?

The contemporary Spherical Great Circle Diameter on the Second Dimensional Plane X **(½ pi) 1.570795** = the Laviolетteian Superior Circle Diameter. **2 * 1.570795 = 3.14159**.

Contemporary Great Circle

πr^2 x 4 Great Circles

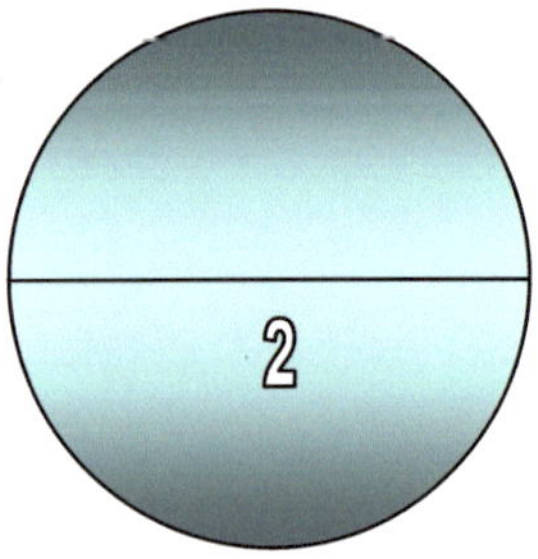

Laviolетteian Superior Circle

πr^2 x 2 Superior Circles

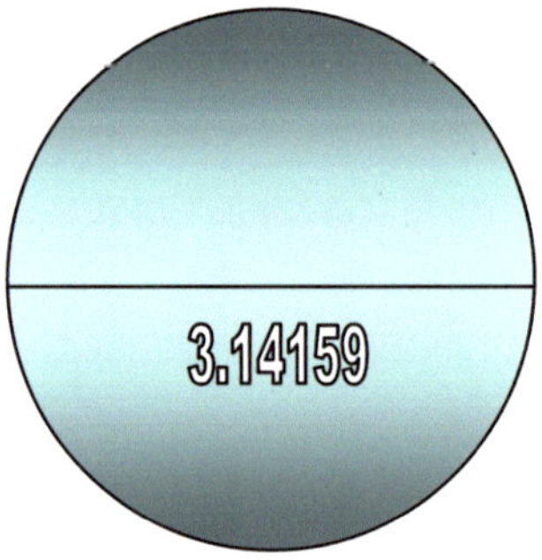

Contemporary
12.566358

SECOND DIMENSIONAL
SPHERICAL SURFACE AREA

Laviolетteian
15.503098

Laviolетteian Surface Area divided by Contemporary Surface Area reveals .2336985 % more Second Dimensional Spherical Surface Area. **15.503098 / 12.566358 = 1.2336985 %.**

If the Contemporary Surface Area is multiplied by **(½ pi),** the new Contemporary Surface Area is **12.566358 x 1.570795 = 19.739172**.

If the Laviolетteian Surface Area is divided into the new Contemporary Surface Area it will reveal the percent amount of additional Laviolетteian material needed to equal the new Contemporary Surface Area.

19.739172 / 15.503098 = 1.2732404
x 1.2732404
19.73917

SEE CONSTRUCTIVE DEMONSTRATION PAGE 14:

6) 27.32405% material would have to be added if the square was converted to the circle.

In addition, if the Lavioletteian Surface Area is multiplied by **(½ pi)** it reveals .2336985 % more Second Dimensional Spherical Surface Area. **15.503098 x 1.570795 = 24.352188**

24.352188 / 19.739172 = 1.2336985 %.

Difference between Contemporary missing matter percent .2732405 and Lavioletteian Superior Circle Diameter increase of **(½ pi) 1.570795** reveals the .2336985 % of missing matter called **Dark Matter**.
1.570795 / 1.2732405 = 1.2336985 = .2336985 % more Second Dimensional Spherical Surface Area.

THE QUESTION IS:

IS MATERIAL IMMATERIAL RELEVANT TO SUPPLYING MATERIAL TO COVER SURFACE AREA OF A SHAPE ON THE SECOND-DIMENSIONAL PLANE IF PERIMETER OR CIRCUMFERENCE IS ALTERED TO CREATE A DIFFERENT SHAPE?

CONSTRUCTIVE DEMONSTRATION:

GIVEN:

A) Construct an eight (8) inch linear measurement.
B) Convert the linear measurement to create, on a Second-Dimensional Plane, a square, a rectangle and a circle.
C) Calculate the surface area for each conversion.
D) Compare any differences of surface area to determine if material needs to be added or material needs to be discarded if material was supplied to cover the surface area according to the formula yielding the accurate amount of material needed to cover the shape calculated.

THEOREM-L1: Altering perimeter or circumference on Second-Dimensional Plane creates plus or minus factor relevant to discarding or adding material of the original shape to accurately cover surface area of the new shape in accordance with the formula relevant to the new shape if material was utilized and equal perimeter was a maintained factor or the altering creates a Third-Dimensional shape retaining all the material.

PROOF: Mathematically self-evident.

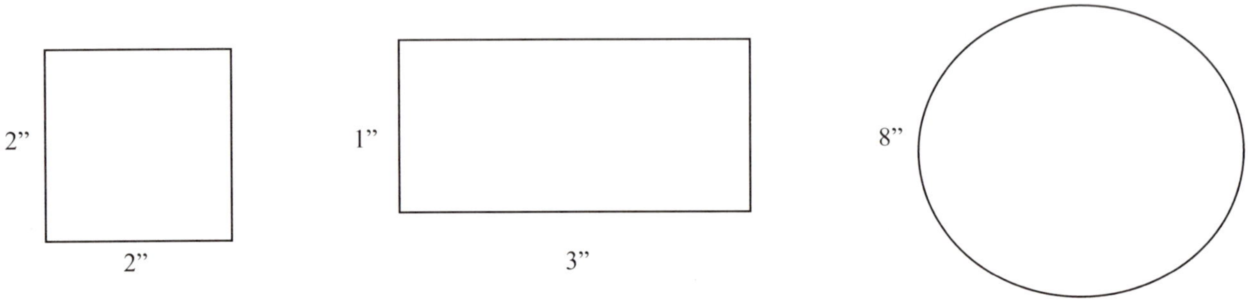

1) Surface area for the square is 4 sq. ins.
2) Surface area for the rectangle is 3 sq. ins.
3) Surface area for the circle is 5.0929622 sq. ins.
4) 25.% of material on hand would be discarded if the square was converted to the rectangle.
5) 33.33333% of material would have to be added if the rectangle was converted to the square.
6) **27.32405% material would have to be added if the square was converted to the circle.**
7) 21.46025% material would have to be discarded if the circle was converted to the square.
8) 41.09519% material would have to be discarded if the circle was converted to a rectangle.
9) 69.7654% material would have to be added if the rectangle was converted to the circle.

CONCLUSION:

Altering shape of material on a Second-Dimensional Plane creates a new shape that is neither symmetrical nor congruent with the original shape. The beauty or harmony of form resulting from a symmetrical or nearly symmetrical arrangement of parts is lacking as is due or right proportion relevant to formulas. Material **<u>IS NOT IMMATERIAL</u>** relevant to calculated surface area expression.

Before we continue with CONTEMPORARY / LAVIOLETTEIAN views of the Second & Third-Dimensional planes; let us first view Contemporary and Lavioletteian calculations for Mass, Source Strength, and Intensity.

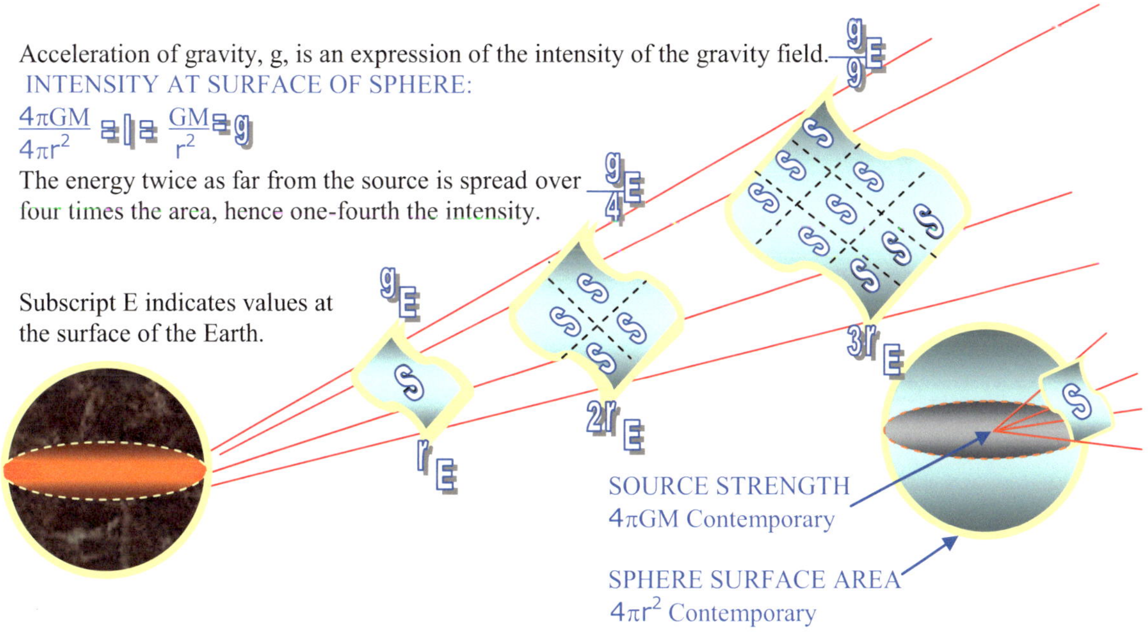

INVERSE SQUARE LAW, GRAVITY

INTENSITY AT SURFACE OF SPHERE:

LAVIOLETTEIAN GEOMETRY UTILIZES $2\pi GM$ AND $2\pi r^2$ FORMULAS.
$2\pi r^2$ IS CALCULATED FROM LAVIOLETTEIAN SUPERIOR CIRCLES.

$$\frac{2\pi GM}{2\pi r^2} = I = \frac{GM}{r^2} = g$$

Other applications of the inverse square law are: LIGHT, SOUND, ELECTRIC FIELD, and RADIATION.

Calculate Mass, Source Strength and Intensity for our **2.528" diameter sphere** on the second-dimensional plane after being converted to LAVIOLETTEIAN SUPERIOR CIRCLES. **Reference page 9 conversion methods.**

2.528" GREAT CIRCLE CONVERTED TO CONTEMPORARY SUPERIOR CIRCLE

MASS	SOURCE STRENGTH	INTENSITY
$M = ar^2/G$-approximation	$SS = 2\pi GM$	$I = 2\pi GM / 2\pi r^2$
		$I = GM / r^2 = g$
M = 9.807 X 3.1953918 / 6.67	SS = 2 X 3.14159 X 6.67 X 4.6982319	I = 31.337206 / 3.1953918
M = 31.337207 / 6.67	SS = 196.8973 = $2\pi GM$	I = 9.8069995 = 9.807 g
M = 4.6982319		
± .0000001%		

2.528" GREAT CIRCLE CONVERTED TO LAVIOLETTEIAN SUPERIOR CIRCLE

MASS	SOURCE STRENGTH	INTENSITY
$M = ar^2/ G$-approximation	$SS = 2\pi GM$	$I = 2\pi GM / 2\pi r^2$
		$I = GM / r^2 = g$
M = 9.807 x 3.9421498 / 6.67	SS = 2 X 3.14159 X 6.67 X 5.7962013	I = 38.660662 / 3.9421498
M = 38.660663 / 6.67	SS = 242.91189 = $2\pi GM$	I = 9.8069997 = 9.807
M = 5.7962013		
± .0000001%		

A Unique Probe of Dark Matter Via Gravitational Lensing

Results of **Einstein's** gravitational lensing calculations validates **Lavioletteian's** geometrical theorem revealing materialization of **DARK MATTER**.

Einstein Ring Calculation

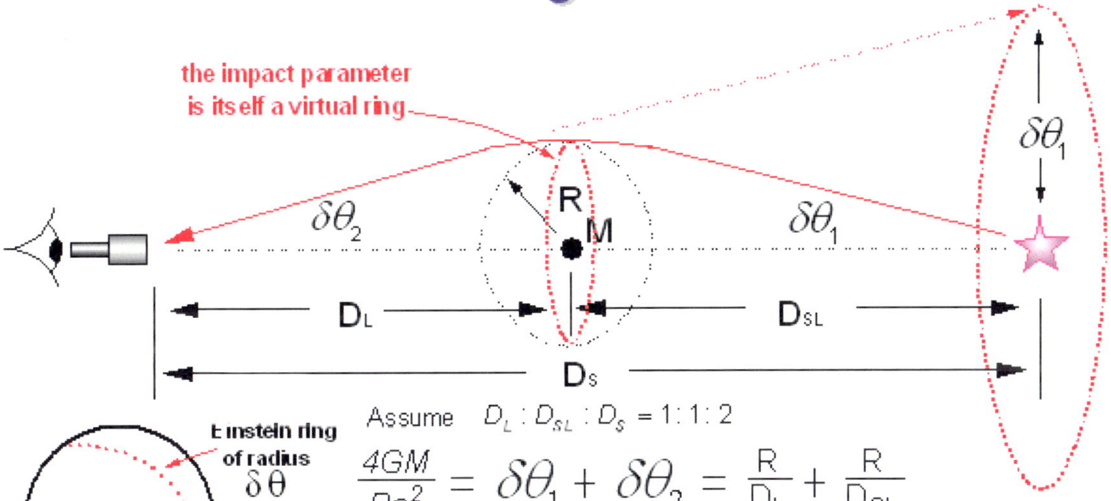

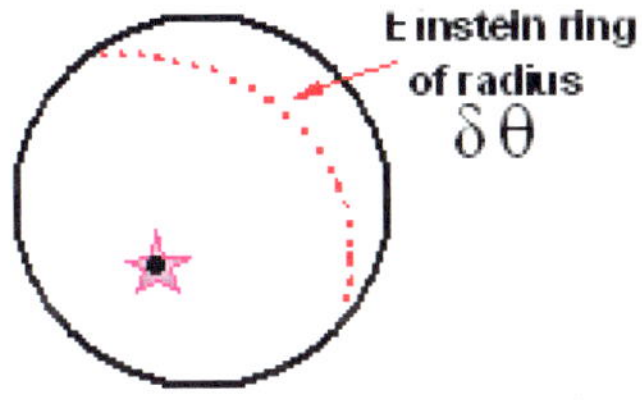

Assume $D_L : D_{SL} : D_S = 1:1:2$

$$\frac{4GM}{Rc^2} = \delta\theta_1 + \delta\theta_2 = \frac{R}{D_L} + \frac{R}{D_{SL}}$$

$$\alpha = \delta\theta_1\Big|_{approach} + \delta\theta_2\Big|_{recede} = \frac{4GM}{Rc^2}$$

$$\frac{4GM}{c^2} = R^2\left(\frac{1}{D_L} + \frac{1}{D_{SL}}\right) = R^2\left(\frac{D_{SL}+D_L}{D_L D_{SL}}\right)$$

$$\delta\theta_{TOTAL} = \frac{R}{D_L} = \sqrt{\frac{4GM}{c^2 D_L}\frac{D_{SL}}{D_{SL}+D_L}}\Bigg|_{\text{general case}} \qquad \delta\theta = \sqrt{\frac{2GM}{c^2 D_L}}\Bigg|_{\text{special case } D_L = D_{SL}}$$

LAVIOLETTEIAN RADIUS

THEOREM-L2: Two SUPERIOR CIRCLES having diameters equal to 180 degrees of a **Third-Dimensional** spherical circumference yields true surface area and matter distribution of a sphere on **the Second-Dimensional plane**. Surface area is 23.36985 percent more than contemporary calculation of four GREAT CIRCLES on **the Second-Dimensional plane**. Reference page 6.

According to our current understanding, a star and its planets form out of a collapsing cloud of dust and gas within a larger cloud called a nebula. The nebula condensed and became a spinning disk.

These **"circumstellar" or "protoplanetary" disks**, as astronomers call them, are the birthplaces of planets.

Can we not visualize the disk below as being **Einstein's Ring** on a 2D flat surface?
Can we not visualize this disk as being **Two Lavioletteian Superior Circle**s on a 2D flat surface before collapsing into a sphere? Both methods embrace all the matter in a disc acknowledging **DEPTH FACTORS.**

Archimedes concern was surface area of a sphere on a Second Dimensional Plane; not **DEPTH FACTORS** transferring all the matter of a disc to a sphere as we do when calculating total mass.

D = 3.14159 divided by (½ pi) 1.570795 equals D = 2

SUPERIOR CIRCLE HEMISPHERE

Laviolette's Superior Circles are drafted from <u>SETTLED SPHERICAL DIAMETERS</u>.

As the hemisphere expands outward to create the LAVIOLETTEIAN SUPERIOR CIRCLE the Second Dimensional **<u>circular diameter will be equal to the 180 degree arc of the hemisphere</u>.** Therefore, one could visualize a **<u>Third Dimensional</u>** figure due to **<u>DEPTH</u>** created from **<u>the settled matter coming from the sphere</u>** or rationalize its appearance as a nebula disk having depth. However, we calculate Second Dimensional surface area from the bird's-eye view of the **<u>HEMISPHERICAL OR SPHERICAL SETTLED MATTER</u>.**

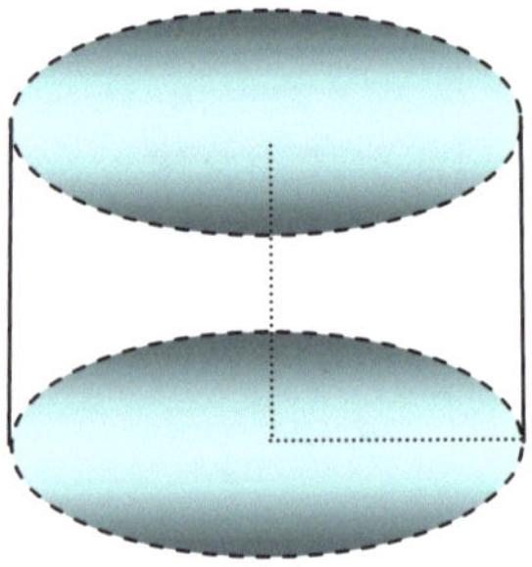

All Nebula Disk Have Depth

When the Sphere is converted to Laviolettеian Superior Circles its settled diameters are the correct diameters to calculate Surface Area of that Third-Dimensional Sphere.

Disk in **Nebula picture** represents Third-Dimension unsettled matter, **<u>having depth</u>,** before gravity begins to pull the disk towards its own center of mass until the disk collapses into a sphere retaining all the settled matter for that sphere after **Orbital Clearing.** If you visualize the disk in the picture above as a converted hemisphere into a Laviolettеian Superior Circle you can look down upon it to see and calculate Third-Dimension Spherical Surface Area on a Second-Dimensional Plane. However, the Laviolettеian Superior Circle's diameter can never exceed ½ of the targeted <u>settled Spherical Circumference</u> after the spherical collapse. A sphere's two Superior Circles on a Second Dimensional plane reveals **<u>total matter</u>** of the sphere <u>having no depth on a Second Dimensional plane</u>.

Since the weights of different bodies at the same location are proportional to their masses, weight is often used as a measure of mass. Therefore, if you assign a weight to the square units of Superior Circles on the Second Dimensional Plane you not only account for the 23.36985 % of missing matter, relevant to the weight factor, but its distribution as well. In addition, when the Superior Circles arc back to a **Third-Dimensional** Sphere the matter is redistributed to the **<u>surface and ball</u>** of the sphere **as PERCENT FACTORS. 52.62281% of the additional matter is allocated to the ball of the sphere (23.36985 % - 11.07198 % = 12.29787 %).**

DARK MATTER

If there is enough matter to close the universe gravitationally then it is wrapped in on itself like a sphere. If there isn't enough matter to close the cosmos then our universe has an open shape extending forever in all directions. Unlike closed universes, open universes have an infinite amount of space.

Matter has important consequences for the evolution of the Universe and its structure within. According to general relativity, the Universe must conform to one of three types: open, flat, or closed. The total amount of mass and energy in the Universe determines which of the three possibilities applies to the Universe. The type of matter, **"Dark Matter or Baryonic Matter"**, will have a profound effect relevant to an open, flat, or closed Universe.

Scientists still do not know if there is enough matter in the universe to close it. If it is closed what is the hidden matter closing it?

Science, for the first time is asking the public to accept something they cannot find, produce or demonstrate. They want us to accept 23 percent of **DARK MATTER** because they are sure it's needed to explain the universe. The effects of gravitational pull confirm that the dark matter or something must pervade our Universe.

What we know for sure is the fact we need at least 23 % more matter and it must be composed of ATOMS. **The 23 % of more matter composed of atoms is all we need to prove because it is that percentage that kept the computerized virtual galaxies from falling apart.**

The proof of the nonexistence of dark matter besides not being able to see it with any wavelength of the electromagnetic spectrum, besides not being able to collect and hold in one's own hand, besides not being able to detect or create dark matter in any laboratory setting should be proof enough that dark matter is nonexistent. **PROOF** of the nonexistence lies with a glaring procedural mathematical error. Revelation of this truth will not only reveal **23.36985 % more "Baryonic Matter"** but will account for the motions of astronomical objects, specifically stellar, galactic, and galaxy cluster/supercluster observations as well. It will also explain why **velocity measurements** done on large scales reveal much more mass than can be explained by the luminous stuff.

IS IT JUST A COINCIDENT A PROCEDURAL MATHEMATICAL ERROR REVEALS 23.36985 % MORE "BARYONIC MATTER" and that Dark Matter inferred a 23 % increase of an unknown source of matter?

THE SCIENTIFIC COMMUNITY STATES:

1). There is 23 percent more matter. Dark Matter fills the void. But, it cannot be detected. Its presence is inferred indirectly from the motions of astronomical objects. By measuring these mysterious effects of gravity research determined how much "EXTRA" matter must exist.

2). Newton's Laws may need to be modified. Dr. Vera Rubin at the Carnegie Institution worked with the outside of galaxies. She discovered stars further out from the center were moving just as fast as those near the center. The stars seem to defy the laws of gravity. Rubin said it was expected the velocity of the stars would fall off but they were straight out. It was as though they were being held in place by extra gravity.

3). Einstein's relativity is more accurate relevant to gravity.

WHAT IS THE PROBLEM?

Science detected extra gravity and ASSUMED extra matter must exist. The extra amount of matter is 23 percent rounded off. Some scientists believe the extra matter is there in the form of "DARK UNDETECTED MATTER" because solar systems or galaxies are not flying apart. **Laviolette states Dark Matter does not exist relevant to its intended purpose or for that matter any purpose.**

Geometrical experiments reveal 23.36985 % more Spherical Surface Area on the Second-Dimensional Plane in opposition to contemporary belief.

Mathematical discovery corrects a geometrical error intrinsic in the work of Archimedes.

What is the nature of "DARK MATTER", this mysterious material that exerts a gravitational pull, but does not emit nor absorb light? Astronomers do not know.

Dark Matter is not made of atoms**?** Scientist are looking for something that has mass but isn't made of atoms. Something that is everywhere but cannot be seen. Something that exerts a gravitational pull, but does not emit nor absorb light.

New results will dramatically change the scientific perception of our Universe.

GEOMETRY:
HOW ACCURATE IS IT?

I BELIEVE THE FLAW I DISCOVERED IN GEOMETRY WILL MATERIALIZE
DARK MATTER.

THE FLAW HAS MASS BUT ISN'T MADE OF ATOMS.
THE FLAW IS EVERYWHERE AND HAS NOT BEEN SEEN.
THE FLAW EXERTS A GRAVITATIONAL PULL, BUT DOES NOT EMIT NOR ABSORB LIGHT.

The contemporary Second Dimensional Plane when utilized to reveal Surface Area of a **Third-Dimensional** Sphere via a **formula** must provide SUFFICIENT **Second-Dimensional** SURFACE COVERAGE to account for natures **Accretion Factor** if it were activated to reveal settled matter of a **Third-Dimensional** Sphere from a hypothetical Second Dimensional Plane.

FRITZ ZWICKY

The father of Dark Matter

Fritz Zwicky was the first to recognize that in rich clusters of galaxies, a large portion of the matter is not visible. In his pioneering work he estimated the total mass of the COMA cluster of galaxies from the motions of the galaxies within that cluster. Using the **virial theorem** to relate the total average kinetic energy and total average potential energy of the galaxies of the Coma cluster he came to the conclusion that the galaxies were on average moving too fast for the COMA cluster to be held together only by the mass of the visible matter. He coined the phrase **Dark Matter.**

Laviolette states:

MATHEMATICAL ERROR REVEALS DARK MATTER IS NONEXISTENT.
Any formula utilizing $4\pi r^2$ or any multiple of the formula understates its universal worth. All the $4\pi r^2$ formulas below understate value of the equations by .2336985 % and Third Dimension Spherical Surface Area is understated by.1107198 %.

VIRIAL THEOREM EQUATIONS:

THERMAL (KINETIC) ENERGY: $KE = \int_0^R \frac{3}{2} KT * N * 4\pi r^2 dr$

GRAVITY POTENTIAL ENERGY: $PE = -\int_0^R \frac{GM}{r} P\, 4\pi r^2 dr$

IDEAL GAS LAW $P = NKT$ so: $KE = \int_0^R \frac{3}{2} P(r) * 4\pi r^2 dr$

INVERSE SQUARE LAW EQUATIONS:

ELECTRIC: $E = \frac{Q}{4\pi \varepsilon_0 r^2}$

RADIATION: $\frac{S}{4\pi r^2}$

GRAVITY: $I = \frac{GM}{r^2} = g$

SPHERICAL ILLUMINATION: Surface area is understated by .1107198 %.

16

Fritz Zwicky never conceived the thought that surface area of a sphere was misrepresented on the Second-Dimensional plane relevant to its **FORMULA** and the fact that the misrepresentation revealed a **.2336985 % loss** when expressed as $4\pi r^2$. The loss should have been added to the conventional $4\pi r^2$ formula evident when expressed as Superior Circles on the Second-Dimensional Plane. **Example:** $4\pi r^2 =$ **12.566358 x 1.2336985 = 15.503098.** **See page 8.**

LAVIOLETTEIAN FORMULS:

SURFACE AREA OF A SPHERE ON SECOND-DIMENSIONAL PLANE IS:

NEW FORMULA (MOST ACCURATE) **.25C=r**
($2\pi r^2$ = SA)

SURFACE AREA OF A SPHERE ON THIRD-DIMENSIONAL PLANE IS:

NEW FORMULA (MOST ACCURATE) **.25C=r**
($2\pi r^2$ x .900317= SA)

The observable universe appears from our perspective to be spherical. We assume the Universe is isotropic. The distance to the edge of the observable universe is roughly the same in every direction—that is, the observable universe is a spherical volume (a ball) centered on the observer, regardless of the shape of the Universe as a whole. **In addition, we calculate clusters and super-clusters spherically when weighing the Universe.** Therefore, Laviolettеian Second-Dimensional plane, representing surface area of a sphere, **REVEALS** 23.36985 % **more** universal missing matter referred to as **DARK MATTER**. It **VALIDATES** 23.36985 % **more** Force, 23.36985 % **more** mass, 23.36985 % **more** weight, 23.36985 % **more** source strength, 23.36985 % **more** resistance known to the scientific community and **REVEALS** 11.07198 % **more** Third-Dimensional Spherical Surface Area. Also, 52.62281% of the additional matter is allocated to the ball of the sphere (23.36985 % - 11.07198 % = 12.29787 %).

Mathematics must be calculated from a source and its quantity, measurement or dimension rather than an approximation of an assumption relevant to proportionality.
Eugene J. Laviolette

The following represents Lavioletteian percentage of redistribution for the standard model of the Universe.

CONTEMPORARY DISTRIBUTION

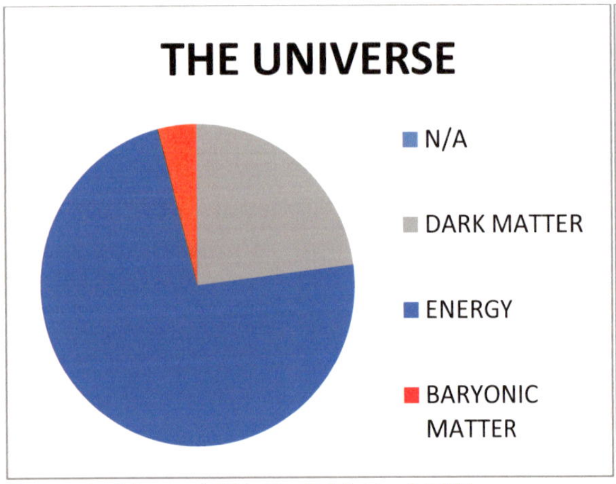

Baryonic Matter 4.9%, Dark Matter 26.8%, Energy 68.3%

LAVIOLETTEIAN REDISTRIBUTION

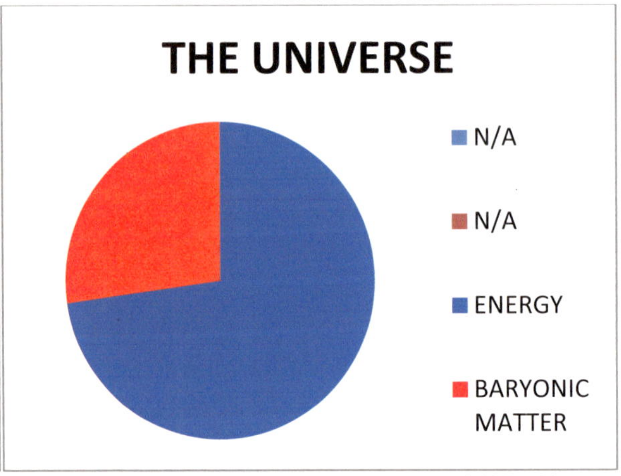

Baryonic Matter 4.9% + 23.36985% = 28.26985 %, Energy 71.73015%

THE UNIVERSE
71.73015 % energy + 28.26985 % Baryonic Matter = 100 %
All of our Universal calculations are Spherical in nature.

CONCLUSIONS:

1.) Lavioletteian Second-Dimensional circular surface area is 23.36985 % **MORE** surface area than Contemporary Second-Dimensional circular surface area.

2.) Lavioletteian Third-Dimensional spherical surface area is 11.07198 % **MORE** surface area than Contemporary Third-Dimensional spherical surface area.

3.) 15.503098 Lavioletteian Second-Dimensional circular surface area.
x .900317 %
13.957702 = Lavioletteian Third-Dimensional spherical surface area.

13.957702 / 12.566358 Contemporary Third-Dimensional spherical surface area = 1.1107197 %.
.1107197 % **REVEALS more** Lavioletteian Third-Dimensional Spherical Surface Area compared to Contemporary Third-Dimensional spherical surface area.

4.) 13.957702 Lavioletteian Third-Dimensional spherical surface area / 15.503098 Lavioletteian Second-Dimensional circular surface area = .9003169
.9003169
-1.0000000
.0996831 % **Total** Lavioletteian Second-Dimensional circular surface **area percentage allocated to the ball of the** Lavioletteian Third-Dimensional **sphere?** Reference page 11.

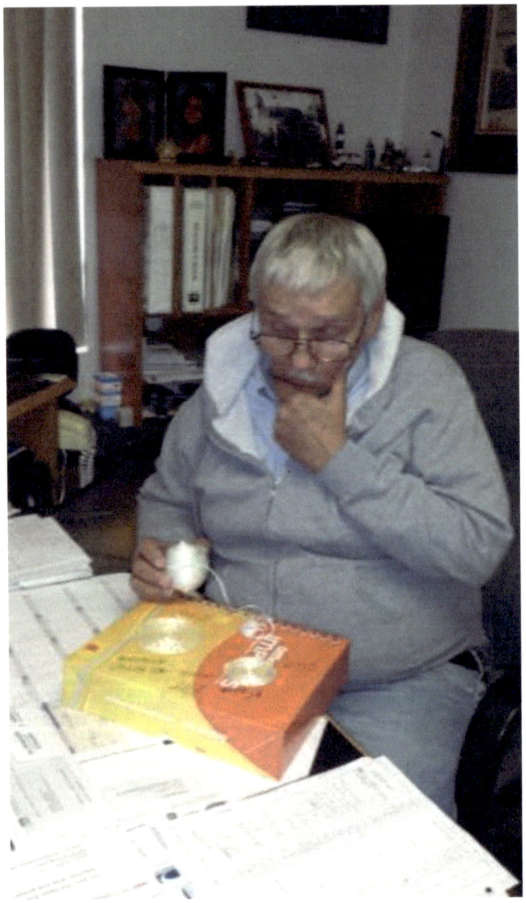 

Author Eugene J. Laviolette Now and Then

In mathematics, a careful distinction is made between the sphere (a two-dimensional surface embedded in three-dimensional Euclidean space) and the ball (the interior of the three-dimensional sphere).

Scientists are working with (a two-dimensional surface embedded in Three-Dimensional Euclidean space) rather than arcing Two-Dimensional Euclidean circles, referred to as Laviolетteian Superior Circles, into hemispheres which demonstrate correct and accurate distribution of Euclidean Second-Dimensional matter which reveals 23.36985 % additional **BARYONIC MATTER** referred to as ***inferred*** **DARK MATTER** while maintaining the Great Circle of the Sphere from which the Laviolетteian Superior Circles were drafted.

THE SPHERE

When you look at a sphere such as the sun or a heavenly body you are unaware of the following facts:
1) The sphere has **23.36985 %** more mass known as **"Baryonic Matter".**
2) Surface Area is **11.07198 %** more **Baryonic Matter** than contemporary belief.

Made in the USA
Las Vegas, NV
13 November 2026

58093850R10019